X-TREME FACTS: ENGINEERING

CITIES

by Catherine C. Finan

BEARPORT PUBLISHING

Minneapolis, Minnesota

Credits:
Cover and title page, Songquan Deng/Shutterstock; 4, Ryan DeBerardinis/Shutterstock.com; 4 bottom left, 5 bottom middle left, 7 top left, 12 bottom right, 20 bottom left, 23 top right, LightField Studios/Shutterstock; 4 bottom right, 7 bottom middle, BearFotos/Shutterstock; 5 top, 26 bottom, 28 top left, ESB Professional/Shutterstock.com; 5 top right, 7 bottom left, 19 bottom left, 24 bottom right, Roman Samborskyi/Shutterstock; 5 middle, Marcin Osman/Shutterstock; 5 middle right, Inside Creative House/Shutterstock; 5 bottom left, mspoli/Shutterstock; 5 bottom right, Mateusz War/Creative Commons; 5 bottom middle right, William Moss/Shutterstock; 6 top, ImageBank4u/Shutterstock; 6 bottom, Louis Darp/Shutterstock.com; 7 top, Pani Garmyder/Shutterstock; 7 top right, MishaShiyanov/Shutterstock; 7 middle, Bobik/Creative Commons; 7 bottom, vitmark/Shutterstock; 8 top, Thomas Lawrence/Birmingham Museum and Art Gallery/Public Domain; 8 bottom, 8 bottom left, 8 bottom right, © The Trustees of the British Museum, released as CC BY-NC-SA 4.0; 9 top, Public Domain; 9 top right, Public Domain;; 9 bottom, William Louis Sonntag, Jr/Public Domain; 9 bottom left, Portrait Image Asia/Shutterstock; 9 bottom right, alexandre zveiger/Shutterstock; 10, Detroit Photographic Company/Public Domain; 10 top right, Evannovostro/Shutterstock.com; 10 bottom right, 10 bottom left, Everett Collection/Shutterstock; 11, T photography/Shutterstock; 11 bottom left,Paul Michael Hughes/Shutterstock; 12, K-digitalstudio/Shutterstock; 13 top, S-F/Shutterstock; 13 top right, iofoto/Shutterstock; 13 middle left, Klintsou Ihar/Shutterstock; 13 middle right, Steven Pisano/Creative Commons; 13 bottom,Craig Mayhew and Robert Simmon, NASA GSFC/Public Domain; 13 bottom right, NASA/Public Domain; 14 top, Benny Marty/Shutterstock.com; 14 bottom, Denis Belitsky/Shutterstock; 15 top, TierneyMJ/Shutterstock; 15 top right, Jeka/Shutterstock; 15 middle, Great Pics - Ben Heine/Shutterstock.com; 15 middle right, Lucy.Brown/Shutterstock.com; 15 bottom, Tomas Marek/Shutterstock; 16 top, jorgeflorez/Shutterstock; 16 bottom, Artens/Shutterstock; 16 bottom middle, ViDI Studio/Shutterstock; 17 top, Sunflowerey/Shutterstock; 17 top right, Lucas Eduardo Benetti/Shutterstock; 17 middle, Puripat Lertpunyaroj/Shutterstock; 17 bottom, Oppdowngalon/Shutterstock.com; 17 bottom middle, Anurak Pongpatimet/Shutterstock; 18 top, Yasonya/Shutterstock; 18 bottom, Efired/Shutterstock; 19 top, gokturk_06/Shutterstock; 19 middle, Kevin Ruck/Shutterstock; 19 bottom, pisaphotography/Shutterstock; 19 bottom right, expresin/Shutterstock.com; 20 top right,A_Sh/Shutterstock; 20 top left, stilrentfoto/Shutterstock; 20 middle, Dmitry Kalinovsky/Shutterstock; 20 bottom, AlexDreamliner/Shutterstock; 20 bottom right, MNBB Studio/Shutterstock; 21 top, Denis Belitsky/Shutterstock; 21 top middle, Goldeyr/Creative Commons; 21 top bottom, Kletr/Shutterstock; 21 middle, Edibobb/Creative Commons; 21 bottom, Belikova Oksana/Shutterstock.com; 22 top, fritz16/Shutterstock.com; 22 top left, Elnur/Shutterstock; 22 top right, Milkovasa/Shutterstock; 22 bottom, Viault/Creative Commons; 22 bottom left, Matyas Rehak/Shutterstock.com; 23 top, Interfase/Public Domain; 23 middle, Homo Cosmicos/Shutterstock; 23 bottom, Rudy Balasko/Shutterstock; 23 bottom right, Philip Schubert/Shutterstock.com; 24 top, thinkhubstudio/Shutterstock; 24 bottom, metamorworks/Shutterstock; 25 top, Natchapol Laotrakul/Shutterstock; 25 top right, cristovao/Shutterstock; 25 middle, FrimuFilms/Shutterstock.com; 25 middle right, 24Novembers/Shutterstock; 25 bottom, Littleaom/Shutterstock.com; 25 bottom left top, Jihan Nafiaa Zahri/Shutterstock; 25 bottom left bottom, Cagkan Sayin/Shutterstock.com; 25 bottom right top, fizkes/Shutterstock; 25 bottom right bottom, Vershinin89/Shutterstock.com; 26 bottom left, Gelpi/Shutterstock; 27 top, Kristin Spalder/Shutterstock.com; 27 top left, paffy/Shutterstock; 27 top right, Brocreative/Shutterstock; 27 middle, swissdrone/Shutterstock; 27 bottom, ArtisticPhoto/Shutterstock; 27 bottom right, Michael C. Gray/Shutterstock; 28 bottom left, Chantal de Bruijne/Shutterstock.com; 28–29, Austen Photography

Bearport Publishing Company Product Development Team
President: Jen Jenson; Director of Product Development: Spencer Brinker; Senior Editor: Allison Juda; Editor: Charly Haley; Associate Editor: Naomi Reich; Senior Designer: Colin O'Dea; Associate Designer: Elena Klinkner; Product Development Assistant: Anita Stasson

Produced for Bearport Publishing by BlueAppleWorks Inc.
Managing Editor for BlueAppleWorks: Melissa McClellan
Art Director: T.J. Choleva
Photo Research: Jane Reid

Library of Congress Cataloging-in-Publication Data

Names: Finan, Catherine C., 1972- author.
Title: Cities / by Catherine C. Finan.
Description: Minneapolis, Minnesota : Bearport Publishing Company, [2023] | Series: X-treme facts. Engineering | Includes bibliographical references and index.
Identifiers: LCCN 2022008048 (print) | LCCN 2022008049 (ebook) | ISBN 9798885091664 (library binding) | ISBN 9798885091732 (paperback) | ISBN 9798885091800 (ebook)
Subjects: LCSH: Cities and towns--Miscellanea--Juvenile literature.
Classification: LCC HT152 .F56 2023 (print) | LCC HT152 (ebook) | DDC 307.76--dc23/eng/20220331
LC record available at https://lccn.loc.gov/2022008048
LC ebook record available at https://lccn.loc.gov/2022008049

For more information, write to Bearport Publishing, 5357 Penn Avenue South, Minneapolis, MN 55419.
Printed in the United States of America.

Contents

In the City

A city is a bustling place. Many people shuffle along side-by-side on their way from home to work or out shopping. Packed with millions of people, cities can be spectacular places. They can also be quite a puzzle for **engineers** who have to find the best ways to construct buildings, streets, and **public transportation** systems. What have they come up with?

There are more than 10,000 cities in the world.

Cities are home to more than 50 percent of Earth's **population**—about 4 billion people!

Cities are often centers for business and entertainment. **Many people go to cities for work and for fun.**

Tokyo, Japan, has the largest population of any city in the world. It is home to about 38 million people.

La Rinconada, Peru, is the highest city in the world. It is nearly 17,000 feet (5,180 m) above **sea level**.

The world's most northern city is Longyearbyen, Norway. The most southern city is Ushuaia, Argentina.

Amazing Ancient Cities

While today's cities are modern feats of engineering, cities have been around for a long time. They started when people first settled together to grow crops. People often built their settlements near rivers, which made it possible to travel by boat to trade goods with other people. Over time, many settlements became thriving cities that grew larger and larger. Let's go back in time to check them out!

The world's oldest known city is Jericho, located in modern-day Israel. It dates back more than 11,000 years.

Ancient cities were built with just people's bare hands and simple tools. Talk about extreme engineering!

Even 2,000 years ago, Rome, Italy, **had a population of 1 million people.**

About 2,500 years ago, Carthage, Tunisia, was **one of the richest cities in the world.**

The ancient city Machu Picchu is high up in Peru's Andes mountains. **It was made of huge stones that were likely rolled up a mountain by hand!**

A Revolution Changes Everything

So, how did these ancient cities transform into the cities of today? Things started changing in England in the mid-1700s. At that time, there was a shift from making goods by hand in homes or small shops to manufacturing goods quickly with machines in factories. This was the beginning of the **Industrial Revolution**. Soon, factories spread to the rest of Europe, to the United States, and beyond. Cities were never the same again!

James Watt invented the steam engine in the late 1700s. This technology provided the power that factory machines needed for the Industrial Revolution.

During the Industrial Revolution, many people moved from **rural** areas to cities.

London, England, went from a population of 1 million in 1800 to 6 million just a decade later, largely because of the Industrial Revolution.

During the Industrial Revolution, **cities in Europe and North America grew the fastest.**

As more and more people moved close together, cities became crowded and dirty. This was bad for people's health.

In the late 1800s and early 1900s, **a Second Industrial Revolution with even faster machines caused cities to grow more.**

The Second Industrial Revolution led to improvements in transportation. People could get around cities quickly and easily in trains and cars.

Reaching for the Sky

The millions of people who moved to quickly growing cities all needed places to live and work. As city populations grew, so did buildings. In the late 1800s, engineers started using lighter **steel** instead of heavier **concrete** to build skyscrapers. These super-tall buildings changed the look of cities forever!

Chicago, Illinois, and New York City were among the first cities to have skyscrapers.

The invention of the passenger elevator in 1853 made people more comfortable in tall buildings. They could reach the highest floors without climbing stairs!

New York City has more skyscrapers than any other city in North America.

As more skyscrapers were built, even more people moved into cities. This meant even *more* skyscrapers had to be built!

City **skylines** changed dramatically after the invention of skyscrapers.

Let There Be Light!

Think of a city at night. You might picture a glowing skyline and lampposts lighting the streets. Today, that's true—but it wasn't always that way. Long ago, nightfall meant cities went dark. People mostly just had dim fires to light the night. But over the years, people have found much better ways to see after sunset. Since the invention of the electric streetlight in the late 1800s, cities have become brighter!

The ancient Romans used lamps filled with vegetable oil to light their streets.

About 2,500 years ago, people in Beijing, China, used bamboo pipes to move volcanic gas to streetlamps. **When the gas was lit, so was Beijing!**

In the early 1800s, cities in Europe and the United States began to use gas-lit streetlights. **These were lit every night and put out each morning.**

The first electric streetlight came from Paris, France, in 1878.

On September 4, 1882, New York City **was lit up with electric streetlights for the first time.**

Many of today's streetlights use **LED** technology. **LED lights save energy and last longer than regular light bulbs.**

Even from space, people can see the glow of brightly lit cities at night.

Cities on the Move

Since cities have lots of people on the move, it's important that they have good transportation systems. Most large cities have subways or trains to help people get around easily. For those traveling in cars, a city's network of streets and highways must be designed to handle traffic and keep things moving. Otherwise, the city would come to a screeching halt!

There are about 180 subway systems throughout the world's cities. Many millions of people use them each day.

Some cities have high-speed trains that allow passengers to travel at more than 150 miles per hour (240 kph). Speedy!

Transportation engineers make sure a city's highways and **interchanges** can safely handle many cars.

Many people hop in a taxi to get around. **In Bangkok, Thailand, open-air taxis called tuk-tuks are popular.**

London's famous double-decker buses allow people to travel around the city while getting an awesome view.

Battling the Heat and Cold

Beyond their buildings, lighting, and transportation systems, some cities must also deal with extreme temperatures. In places where the weather is very hot or very cold, cities need to beat the heat and fight the freeze. From cool roofs to heated sidewalks, clever engineering tackles extreme temperatures head-on!

Some city buildings have white roofs to reflect the sun's heat, keeping the buildings cooler.

The dark pavement of a city's streets **can also be coated with white material to cool them down.**

Just changing the color of pavement can change the temperature of a street by 20 degrees Fahrenheit (6.7 degrees Celsius)!

Plant-covered roofs, called green roofs, help keep things cool by providing shade for buildings and streets.

Reykjavik, Iceland, uses water from hot springs to heat houses. The water also runs through pipes beneath sidewalks to melt snow!

Sapporo, Japan, heats its sidewalks to help get rid of some of the 20 ft (6 m) of snow the city gets each year.

Winning against Water

Extreme temperatures aren't the only force of nature that cities have to handle. Some cities must watch out for water! Most cities are built near some kind of water source so residents can easily get the water they need. But some cities face serious water-related challenges, such as water creeping up onto the shore. Let's see how engineering can help them.

Venice, Italy, is built on water. Huge wooden poles were staked into the ground underwater, and then the buildings were constructed on top.

Venice's famous canals often flood because of rising sea levels. Underwater barriers have been built to control the flooding.

Rotterdam, Netherlands, is 90 percent below sea level. The city has a huge wall called the Maeslant Barrier that can open and close to prevent flooding.

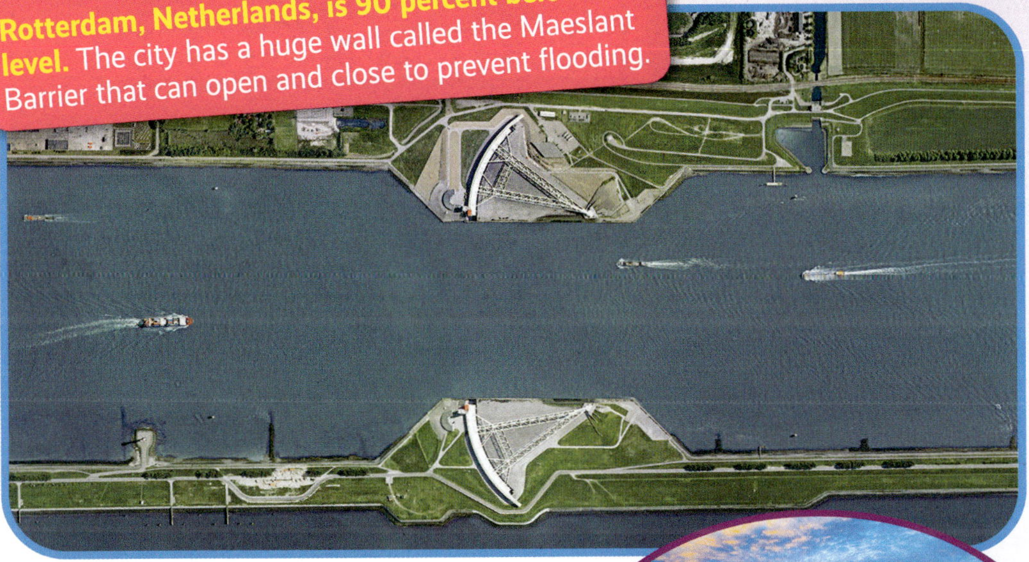

Much of New Orleans, Louisiana, lies below sea level. To help reduce flooding during storms, the city has a huge flood wall that is 1.8 miles (3 km) long.

New Orleans also has parks designed to absorb rainwater.

Because of rising waters around Miami, Florida, engineers want to build new houses on stilts and add more dirt beneath new roads to create higher ground.

Cities in the Clouds

While some cities are built down low, others are sitting high up in the mountains! How have people living at high elevations built spectacular cities? And what do they do about the special problems they face? Let's head to the mountains to learn more!

People can't get to a mountain city without roads! To build these roads, engineers must carefully pick a stable spot on the mountain.

After finding a great spot for a mountain road, **construction begins with blasting away rock to make room for the paved street.**

Heavy machinery is often needed to dig into the frozen ground high up on mountains.

Even in the mountains, engineers have found ways to clean wastewater, which contains people's pee and poop.

The highest up city in the United States is Leadville, Colorado. It's 10,152 ft (3,094 m) above sea level.

La Paz, Bolivia, is a dizzying 11,942 ft (3,640 m) above sea level. **It has the world's longest city cable car system.**

Mind-Boggling Cities

Each city is different in its own way. But some are so unique it's hard to believe they could actually exist! From one city built underground to another floating on a sea, these places show us how extreme city engineering can be! Let's check out some of the wildest cities ever built.

Coober Pedy, Australia, was built inside old **mines. People there live underground, away from the area's extreme heat.**

Iquito, Peru, is the largest city on Earth without road access. With the Amazon River on one side and a rain forest on the other, everything must be brought in by plane or boat.

Neft Daslari is a city of floating metal on the Caspian Sea. It was built as a site for extracting oil from underground.

Neft Deslari has about 185 miles (300 km) of roads.

Siwa Oasis was built in the Sahara desert more than 12,000 years ago. High temperatures there can reach 119°F (48°C)!

Perth, Australia, has 2 million people living in a **remote** location between the Indian Ocean and the unsettled Australian **Outback**.

A Smart City?

You've heard of a smartphone, but what about a smart city? Someday, smart cities could use advanced technology to improve people's lives. How? They would be constantly learning and changing. Technology throughout smart cities could adjust buildings and public transportation as people need them for different things. While it may seem like something out of science fiction, many places already have some smart city features! Let's take a look . . .

A smart city would use a network of sensors and software connecting things to the internet and to each other.

Smart cities could reduce traffic by watching it and **making real-time adjustments to stoplights and public transportation routes.**

Smart city buildings can turn on and off lighting, heating, and cooling as needed. This saves energy.

Boston, Massachusetts, uses cameras and sensors to track traffic patterns in an attempt to reduce car crashes.

In Barcelona, Spain, city garbage bins have sensors that tell them when to suck trash into underground tanks. **This keeps the city from getting stinky!**

Cities of the Future

As more people move to cities, these busy places will continue to grow. We'll need new technologies and engineering to ensure the world's cities can allow people to live healthy, happy lives. Engineers are also working to make cities better for the planet. This could all be achieved by making smart choices in how cities are constructed. Will you be one of the engineers who helps make a difference?

By 2050, it's likely that 68 percent of Earth's population—6.3 billion people—will live in cities. *Whoa!*

Cities with more than 10 million people are called megacities. By 2030, there will probably be more than 40 megacities around the world.

Tomorrow's skyscrapers might be made of wood! Because wood is renewable, it's considered a more environmentally friendly choice than steel and concrete.

Cities of the future could use more solar power and other **renewable resources**.

Many cities are looking at ways to reuse old buildings that would otherwise be torn down. **This helps the cities and the environment.**

Someday, more cities may float on water! These sea cities would make their own energy and food.

Design a Downtown

Craft Project

A lot goes into planning and building a city! There must be places to live, work, and play. Cities also need places that provide the food and services people need. Well-designed streets and transportation should link all these things together. Now, it's time for you to plan your own city's downtown. What will it look like? How will it make the future better for everyone living there? It's up to you!

A downtown is usually near the center of a city and has places for work and entertainment.

Downtowns should be safe and easy for people to walk around.

What You Will Need

- Colored foam board
- White masking tape
- Construction paper
- Scissors
- Tape

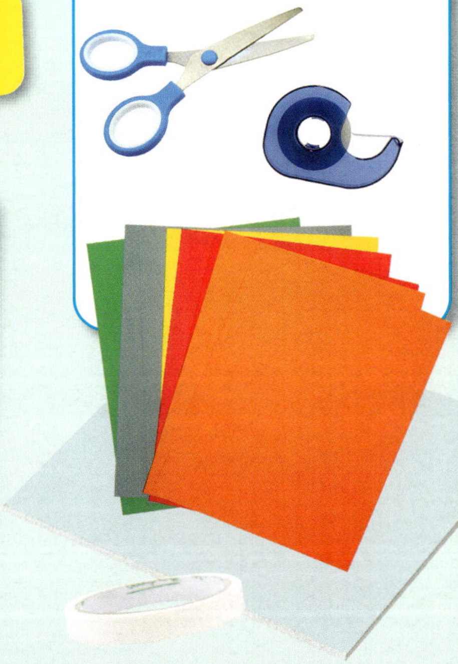

Step One

Use masking tape to make an evenly spaced grid of streets on your foam board. Cut a circle of green construction paper. Tape it to the center of the board. This is a park.

Step Two

Cut pieces of construction paper into different sizes. Fold each piece into a rectangular tube, taping the sides together. These are your buildings. Make many buildings in different sizes.

Step Three

Draw doors and windows on the sides of the buildings. You can draw different things to make them look like different places, such as stores, apartments, or offices. Then, cut pieces of construction paper a little bigger than the tops of each building. These will be their roofs. Use tape to attach the roofs to the buildings.

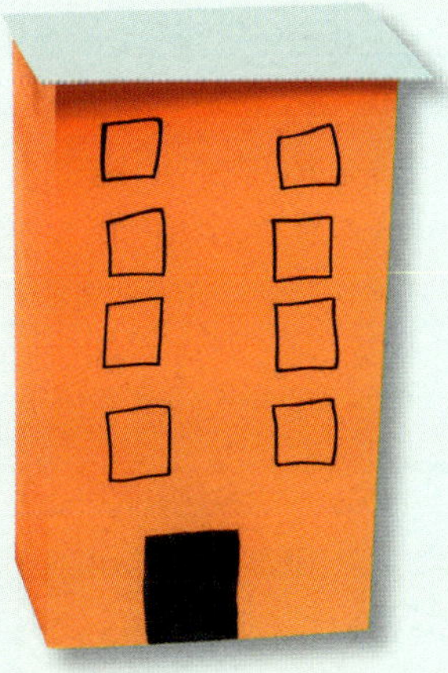

Step Four

Arrange the buildings on the grid. Attach them with tape. Your downtown looks great!

Glossary

canals human-made channels for boat traffic

concrete a hard, strong building material made from sand, gravel, cement, and water

engineers people who use science and math concepts to design buildings, bridges, and roads

Industrial Revolution a period in the late 1700s and early 1800s when people began to build factories and produce a large amount of goods

interchanges places on a highway where different road levels are used to allow vehicles to move from one road to another without disturbing the flow of traffic

LED light-emitting diode; this technology can be made into light bulbs

mines holes and tunnels dug in Earth so people can reach valuable minerals

Outback a large rural area of Australia

population the people who live in an area

public transportation a system of trains, buses, and other vehicles that is operated by the government

remote far away from other cities or towns

renewable resources things that are naturally replenished, or renewed, such as sun and wind

rural having to do with the countryside

sea level the surface level of the sea; this point is often used to measure elevation and depth.

skylines the outlines of natural or human-made structures against the sky

steel a hard, strong metal made from iron mixed with carbon

Read More

Gitlin, Martin. *Smart Cities (Exploring the Internet of Things).* Ann Arbor, MI: Cherry Lake Publishing, 2020.

Klepeis, Alicia Z. *Explore New York (Great Cities of the World).* Mankato, MN: 12-Story Library, 2020.

Kurtz, Kevin. *The Future of Cities (Future Tech).* Minneapolis: Lerner Publications, 2021.

Learn More Online

1. Go to **www.factsurfer.com** or scan the QR code below.
2. Enter "**X-Treme Cities**" into the search box.
3. Click on the cover of this book to see a list of websites.

Index

About the Author

Catherine C. Finan is a writer living in northeastern Pennsylvania. She was once in the highest up city in the United States—Leadville, Colorado—on the Fourth of July. It was snowing!